Danny and the Monarch Butterfly

written by Mia Coulton

photographed by
Mia Coulton and Margaret Ransohoff

It was a hot summer day.

Danny was looking for butterflies in the garden.

Danny found a butterfly sitting on a milkweed leaf.

The butterfly flew away.

Danny looked and looked for the butterfly.
He did not find the butterfly.
But he did find a small white bump
on the milkweed leaf.

It was a butterfly egg.

7

Danny went to look at the egg every day.

After a few days, the egg hatched.
Out came a little caterpillar.

The little caterpillar was hungry.
The little caterpillar ate and ate
the milkweed leaves.

Soon the little caterpillar became a big caterpillar.

The big caterpillar ate and ate the milkweed leaves day after day.

One day, Danny came to look
at the big caterpillar.
It was hanging upside down.

The big caterpillar was very still.
Danny was very still too.

Danny looked at the hanging caterpillar.

It began to change.

Something was growing around the caterpillar.

It looked like a soft blanket.

Danny watched as the soft blanket became small and hard.

It was now a green shell with gold dots.

1, 2 ,3 ,4, 5, 6, 7, 8, 9...

10 days later, the shell changed its color.

It was almost black.

Then the shell changed again.

Now, Danny could see inside the shell.

He saw white dots

and orange and black lines.

All of a sudden, the shell split open.
Something was coming out.

Danny first saw
a wet head,
then wet legs,
then a wet monarch butterfly.

The monarch butterfly dropped
onto Danny's head.
It sat on Danny's soft fur
waiting for its wings to dry.

When its wings are dry,
Danny will have to say goodbye
to his new friend, the monarch butterfly.

Life Cycle

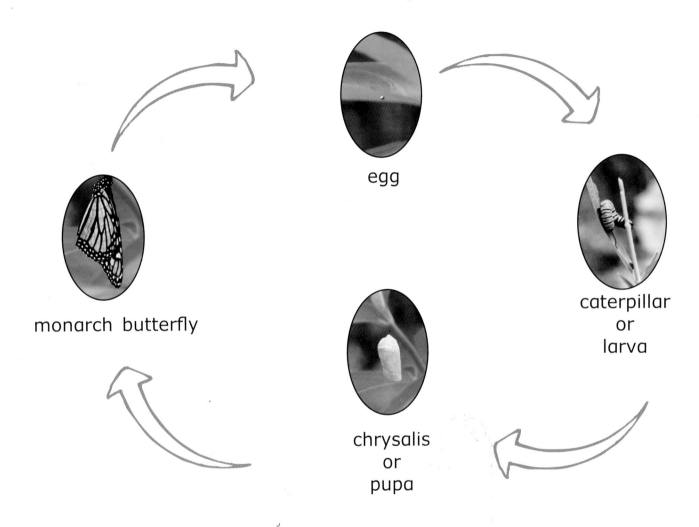

egg

caterpillar
or
larva

chrysalis
or
pupa

monarch butterfly

*When a caterpillar turns into a butterfly this is called **metamorphosis**.*